The D7 Tractor

A VISUAL HISTORY OF THE D7 TRACTOR IN U.S. ARMY SERVICE 1941-1953

by David Doyle

Published by
The Ampersand Group, Inc.
235 NE 6th Ave., Suite B
Delray Beach, FL 33483-5543
561-266-9686 • 561-26609786 Fax
www.ampersandpubco.com

Acknowledgements:
This book would not have been possible without the considerable help of Tom Kailbourn, Jeff Rowsam, Pat Stansell, Eric Reinert, Scott Taylor, Pip Biddlecombe, R. G. LeTourneau, the late David Harper and the late Kevin Kronlund.

Sources:

History of the Development of Mechanical Equipment IX Tractors, Dozers and Their Armor, ME-9 by The Engineer Board, Fort Belvoir, VA

TM 5-3068, Tractor, Crawler, Diesel, 70 to 90-DBHP, Standard, Caterpillar D7, May 1943

TM 5-3068, Tractor, Crawler, Diesel, 70 to 90-DBHP, Standard, Caterpillar D7, March 1945

TM 5-3068, Tractor, Crawler, Diesel, 70 to 90-DBHP, Standard, Caterpillar D7, June 1949

TM 5-3072, Tractor, Crawler, Diesel, 70 to 90-DBHP, Standard, Caterpillar D7, May 1943

TM 5-3073, Tractor, Crawler, Diesel, 70 to 90-DBHP, Standard, Caterpillar D7, November 1944

TM 5-3074, Tractor, Crawler, Diesel, 70 to 90-DBHP, Standard, Caterpillar D7, May 1943

TM 5-3076, Tractor, Crawler, Diesel, 70 to 90-DBHP, Standard, Caterpillar D7, December 1942

TM 5-3086, Tractor, Crawler, Diesel, 70 to 90-DBHP, Standard, Caterpillar D7, January 1944

TM 5-3092, Tractor, Crawler, Diesel, 70 to 90-DBHP, Standard, Caterpillar D7, March 1944

TM 5-3098, Tractor, Crawler, Diesel, 70 to 90-DBHP, Standard, Caterpillar D7, March 1944

TM 5-3108, Tractor, Crawler, Diesel, 70 to 90-DBHP, Standard, Caterpillar D7, May 1944

Photo Credits: U.S. Army Engineer School History Office (USAESHO), U.S. Army Corps of Engineers (USACOE), National Archives and Records Administration (NARA), Military History Institute (MHI), Imperial War Museum (IWM), David Doyle (DD), Pat Stansell (PS), R. G. LeTourneau, Patton Museum and Jeff Rowsam.

Front cover: When one thinks of crawler tractors in the U.S. service from World War II, the model that usually springs to mind is the Caterpillar D7. Part of the U.S. inventory of heavy tractors, the D7 was present on all fronts during the war.

Title page: This D7 is clearing the way for a gasoline-pumping station in the Apennine Mountains in 1945. Blades were not an integral part of the tractor, but rather were accessories, and this D7 features a cable-operated LeTourneau operated via a rear-mounted R7 power control unit. (National Archives)

Rear cover: A restored Caterpillar D7, 7M Series, pauses during a demonstration of its bulldozing capabilities. The addition of LeTourneau bulldozer kits and LeTourneau power-control units gave these already versatile tractors a wealth of additional applications. (DD)

Introduction

The U.S. military has had a relationship with the Caterpillar Tractor Company—Caterpillar, Inc.—since the latter was formed. Cat, as the company is popularly known, was formed in 1925 by the merger of the C.L. Best Gas Tractor Company and the Holt Manufacturing Company, the latter of which used "Caterpillar" as a trade name for its tractors. Holt had supplied full-tracked tractors to the U.S. Army during World War One.

While in addition to Caterpillar, Allis-Chalmers and International Harvester supplied large numbers of track-laying tractors to the U.S. and Allies during WW2, Caterpillar was the primary supplier, and the D7 was the primary model. The decision to make the Caterpillar standard for U.S. Engineers was similar to the decision that the CCKW was to be the U.S. primary tactical transport, with Studebaker and International being used for foreign aid and other branches of service—there were some exceptions, but they were rare.

Counter to what many think, most of the D7 (as well as the D4, D6, D8) and the numerous International Harvester and Allis Chalmers tractors purchased by the military during the war were just that—tractors, not bulldozers.

Caterpillar produced the light (10,500 pounds) D4 tractor starting in 1939. Chalked on the LeTourneau Model A4 dozer blade is the D4's federal stock number, 2410-190-0019. The FSN was the precursor to today's national stock number system, and this number is often seen chalked on Caterpillar tractors in vintage photographs. (USAESHO)

It was not until late in the war Caterpillar and after other similar tractor manufacturers, including International Harvester and Allis-Chalmers, offered components such as dozer blades, cable control units, drag winches, hydraulic units and other machine options.

Prior to and during the war, other allied manufacturers like LeTourneau, LaPlant-Choate, Bucyrus Erie, Hyster, Trackson and many others designed and built attachments for tractors. The Army ordered the tractor with attachments that could include combinations of cable controls, dozer blades and winches from these allied manufacturers.

With that thought in mind, we will first consider the tractor.

The D7 is descended from the RD7 9G, which was produced from 1935 through 1940. The "R" in the model number was dropped in 1937. The Caterpillar D8800 four-cylinder Diesel engine, which developed 69 drawbar horsepower and 82 belt horsepower, powered these 9G machines.

U.S. Army Ordnance bought 72 D7 9G series tractors in 1940, and a further 301 D7 in 1942-1943, all for use as artillery prime movers. In this service they were designated (as were 747 Allis-Chalmers HD-10W and 1,795 International Harvester TD-18 tractors bought for the same purpose) Heavy Tractor M1. However, the bulk of wartime D7 production was under the auspices of the U.S. Army Engineers.

In 1940 the 7M series of D7 was introduced. These machines were assigned Caterpillar serial numbers beginning with 7M0002 and continuing through 7M9999, which was produced in 1944. Late in 1944 the 3T prefix was begun to replace the 7M series. There were also a few tractors assembled with 4T prefixes, those being built late in 1945. Caterpillar delivered to the Army 1,150 D7 during 1942, 1,961 in 1943, 7,094 in 1944 and 7,239 in 1945.

The 3T (and 4T) machines were built until 1954, although post-WW2 the styling of the tractors was altered slightly. Scarce among D7 military tractors are the 6T series built for the U.S. Navy beginning in 1945. The 1,054 machines in this series can be distinguished by their six carrier rollers, versus the five carrier rollers used on the 7M, 3T and 4T series.

Caterpillar does not ever use model year designations. Rather, for documentation and support equipment is segregated by the product serial number prefix. Customers, including the U.S. military, are to quote the serial number prefix as a model reference. Small changes or product improvements are referenced in Cat publications by "effective with serial number XXXX." Beginning in the 1950s, when there is a significant group of product improvements or a significant product change the model numbers started to receive a suffix designation such as D7C, D7E, D7F, D7G and so on.

The Board of Engineer Equipment (later the Engineer Board) had begun on 12 July 1937 to study tractors in earnest. Early in the study full-tracked tractors outshone wheeled tractors with troops. Initial 1938 procurement of fully tracked tractors focused on a 4-1/2 ton model produced by the Cleveland Tractor Company, their dozer-equipped Model BG. Once obtained, the 5th Engineer Regiment tested this machine in the fall of 1938. While the BG could move 16-

times as much material within a given time frame than the previously used animal-drawn equipment, the 5th found the machine lacking in reserve power. Cleveland Tractor responded by suggesting their larger 12,000-pound Model DG. The latter model was larger and heavier, but still within the Army-wide 7-1/2 ton load limit for divisional equipment—and most importantly, provided the needed performance. Captain Gilbert E. Linkswiller of the Engineer Board, observing that Allis-Chalmers, International Harvester, Cleveland Tractor and Caterpillar all offered machines in this weight class, recommended adoption of the 7-1/2 ton medium dozer as standard equipment.

Satisfied with the results thus far, the Board drafted military characteristics for 7,200, 15,000 and 30,000-pound tractors. However, the Board further recommended that no orders be placed at that time, as the tractor industry was continuing to improve its designs. Further, Army policy at that time was such that all track-laying vehicles were to be procured through the Ordnance Department, who favored gasoline engines, while the tractor industry favored Diesel engines in the heavier models.

When large-scale procurement was initiated just prior to the U.S. entry into WW2, Diesel-powered tractors were what were ordered because such machines were the type that were available. On 6 January 1944 responsibility for commercial-type track-laying equipment with speeds less than 12 MPH was finally transferred from Ordnance to the Engineers. Indicative of the bizarre situation that existed prior to January 1944, Ordnance was responsible for the procurement, storage and issue of the tractors while the Engineers had those responsibilities for the attachments, such as dozer blades and mounting equipment.

Like most material produced in the United States during the war, production of tractors was closely regulated. Unlike most cars and trucks, however, tractor production for civilian users continued during the war. The tractors and dozers were badly needed to construct the dozens of new airfields, factories, military installations and roads that were supporting the war effort. In April 1942 the War Production Board assigned the military 85 percent of the nation's crawler tractor production. That 85 percent was to be divided not only among the various branches of the U.S. armed services, but was also to include International Aid—chiefly the British. Naturally, the steel shortage that bottlenecked ship, truck and tank production also impacted tractor production. In early 1942, for example, Caterpillar requested 72,422 tons of steel, and were delivered only 47,633 tons.

The WW2 D7 engine was a big 4-cylinder, Naturally Aspirated (NA), Pre-combustion Chambered (PC), low speed 900 or 1,000 Rated RPM, long stroke 80 HP Diesel engine of Caterpillar design. The industrial version of the engine was called a model D8800.

Unlike modern cars, which are started with electrical starter motors, a small two-cylinder gasoline engine started the Diesel of the D7. That two-cylinder engine in turn is started via a hand crank.

The hand crank for the 2-cylinder gasoline engine comes out the side of the radiator housing toward the left side near the track.

The Caterpillar D6 was a medium tractor with an approximate operating weight of 18 tons. Its military designation was G152 Medium Tractor. Initially designated the RD6 upon rollout in 1935, it was redesignated the D6 two years later. A revamped D6 was introduced in 1941 with the 55-horsepower (drawbar) D4600 engine. (USAESHO)

Later tractors (3T's) have a different starting crank location that extends a shaft vertically through the left side of the engine hood. The operator can crank the starting engine while standing on the track, not having to dismount to ground level.

The hand crank is called a "GRIP" in Caterpillar terminology. It is not considered part of the tool group, rather it is part of the gasoline starting engine system. Operators were trained and admonished to never lay it down loose after starting. Loss of the "Grip" and the machine is rendered "non-mission capable."

Regarding Diesel engines, the D7 3T series was replaced by the 17A tractor prefix in 1955. The tractor looks very similar to the WW2 models. The 17A received a newer postwar model D339 engine—still a large 4-cylinder, but with improved technology and higher horsepower and RPM (128 HP at 1,250 RPM). In the early 1960s the D7C was replaced by the D7E with a 48A prefix, still with the large 4-cylinder but now turbocharged D339 engine rated at 180 HP. It was not until the D7F arrived in about 1970 that the D7 received a 6-cylinder engine. The D333 and D333C engine models were again modernized in the 1980s with the engine being upgraded to the famous 3306PC T and again into the late D7G's as a 3306DIT engine.

Dozer blades

The tractor by itself is no more than a prime mover (as used by the Ordnance Department to tow artillery). However, for earthmoving the tractor can tow scrapers, controlled via cables by the tractor operator, or it can be fitted with a dozer blade. As noted earlier, a number of firms produced equipment for use on the D7, with RG LeTourneau, LaPlante-Choate and ultimately Caterpillar being the primary suppliers of dozer blades. The blade push frames attach to the tractor roller frames on each side. The attachment hinge points are correctly called "Trunnions."

Operation of the blade could be controlled two ways, either through hydraulics or mechanically through cables.

A straight dozer blade, such as the LeTourneau WEK7, designed for the D7, without the ability to angle or tilt the blade, limits the versatility of the single machine. The operator had to be very skilled to ditch, backfill or side cast material with a straight blade frame dozer. Although heavier, the angle dozer was more versatile and required less operator skill to accomplish common tasks.

For this reason, at the outset of the war, the angle dozer was thought the best, although in time the tiltdozer came into favor. Further, in the Carolina maneuvers of 1941 the hydraulic blade control units did not acquit themselves well, being prone to breakdown.

The R.G. LeTourneau company supplied the WCK7 model angle dozer found on many of the combat engineer D7 tractors. The "C" shaped push frame allowed the blade to pivot left or right from a center point. The inclined ladder adjustment on the side of the push frame provided five different settings for adjusting the blade side tilt angle as well. The tractor operator could dismount from the operator's seat and mechanically change the dozer blade from straight to left or right and tilt side to side in less than five minutes.

Operation of the WCK7 blade was controlled by a LeTourneau R7 rear cable control unit or a front-mounted LeTourneau AD7 single drum power control unit. The R7 was an extra heavy-duty double drum rear mount unit for use with large capacity Carryall Scrapers when not coupled with the dozer blade operation, whereas the AD7 was for blade operation only.

Comparative testing of the tilt dozer verses the angle dozer began at Thermal, California in October 1942 by the Engineer Board Desert Test Section. The 104-inch wide tiltdozer blade is mounted at right angles to the long axis of the tractor and cannot be angled. It consists of a tilting-type moldboard, push arms, A-frame and control linkages, and weighs about 2 1/2 tons.

The tilt dozer showed definite increase in ease of operation and improved quality of work. Further tests by the 25th Armored Engineer Battalion reinforced these conclusions, and the tilt dozer was recommended for future acquisition in lieu of the angledozers. The aforementioned tests were conducted with model R4 tractors, but tests in July 1943 with the D7 produced similar results, and the LeTourneau A7 tilt dozer blade was tested by the Board at Thermal and by 355th Engineer General Service Regiment. In addition to the LeTourneau model, the La Plant-Choate Model A72 "Adjustdozer" was tested in September 1943.

Many years later, in the 1980s, hydraulics allowed the introduction of the Power Angle Tilt (PAT) feature found on many dozers sold today. The operator adjusts the blade underway from the operator's seat, increasing machine efficiency by many times.

The PAT feature began with the WW2 LeTourneau manual "CK" angle dozer.

Column after column of U.S. Army tractors and bulldozers are assembled at a U.S. Army Engineers base, Supply Depot G-45, Thatcham, England in April 1944 in preparation for the invasion of Normandy. The column to the far left comprises Allis-Chalmers HD10W tractors, which, like the D7, was among the models designated Heavy Tractor M1. The bulldozers in the second column from the left are Caterpillar D7s. (USACOE)

Top left: The Caterpillar D7 heavy tractor entered production around 1940, with the first series being designated 7M. The vehicle was standardized as part of the Heavy Tractor M1 series. This example bears U.S. Army registration number 9199720 on the hood. **Above left:** The same D7 is viewed from the left side with dirt from recent operations piled up on top of the track roller frame, the beam running along the suspension. The tractor has a power-control unit with a double-deck sheave assembly but no provisions for a bulldozer.

Right: Caterpillar D7 9199720 viewed from the front. Stored on the right fender is a packing crate. Chalked above the grille is the federal stock number, 2410-223-1204. The grille comprised seven vertical and two horizontal bars, with expanded-steel mesh. (USAESHO, all)

Top left: Introduced in 1935, the Caterpillar RD7 tractor was powered by the Caterpillar D8800 four-cylinder Diesel engine. The RD7 was redesignated the D7 in late 1937. Here, an RD7 with a front-mounted winch is viewed from the left front in a January 1937 photo. "RD-7" is stamped on the top of the radiator guard. **Top right:** A Caterpillar RD7 tractor is observed from the rear in a 5 January 1937 photograph. On the rear of the chassis frame is an auxiliary high-speed transmission, below which is a military-style tow pintle mounted on a swinging drawbar assembly. This tractor and those like it were used as artillery prime movers. **Above left:** A Caterpillar D7 is displayed in a November 1943 Aberdeen Proving Ground photo. This example mounts street plates on its tracks to protect paved road surfaces from the grinding action of the stock steel tracks. Grousers also were available to mount on the tracks for added traction in soft ground, mud, and snow. **Above right:** The same D7 shown in the preceding photo is viewed from the right rear. Along the top of the track roller frame stretching between the rear of the idler wheel and the front of the drive sprocket are guards for the idler spring and the sprocket bearing. (NARA, all)

A D7 displays a bulldozer kit. Chalked on the dozer bowl and on the toolbox on the left fender is the federal stock number, 2410-233-5749. Affixed to the front of the grille is a perforated steel guard. The curved steel plate on the center of the dozer bowl provides a replaceable wear surface and is also a reinforcement to allow the tractor and dozer to assist in push loading a Carryall scraper from behind. Extra bracing for the front A-frame is in the form of angle irons. (USAESHO)

Left: The front end of a D7 with a bulldozer kit is displayed. On this example, the reinforcing on the upper center of the dozer bowl is in the form of four plates with horizontal welds between them. To the sides of the reinforcing plates are horizontal welds in the bowl. The bowl was the main part of the dozer. The dozer blade was the strip at the bottom. **Right:** A U.S. Army Caterpillar D7, or Heavy Tractor M1, is viewed from the rear. The federal stock number is chalked on the rear plate of the power-control unit. Mounted on top of that unit is a double-deck sheave assembly, containing two vertical and two horizontal sheaves for routing the power-control unit's right and left cables when controlling a bulldozer or operating an accessory. (USAESHO, both)

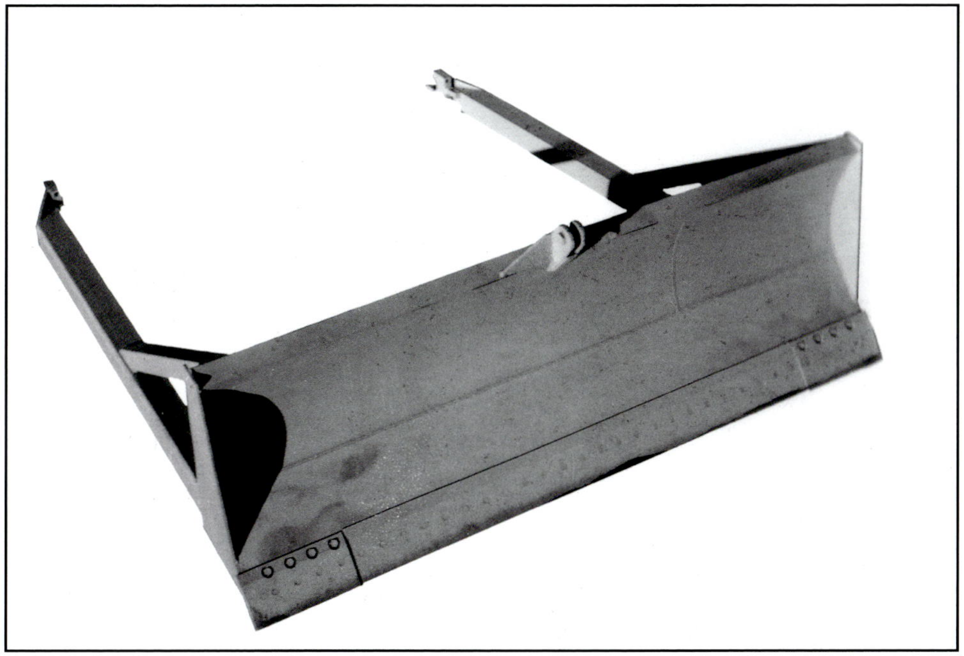

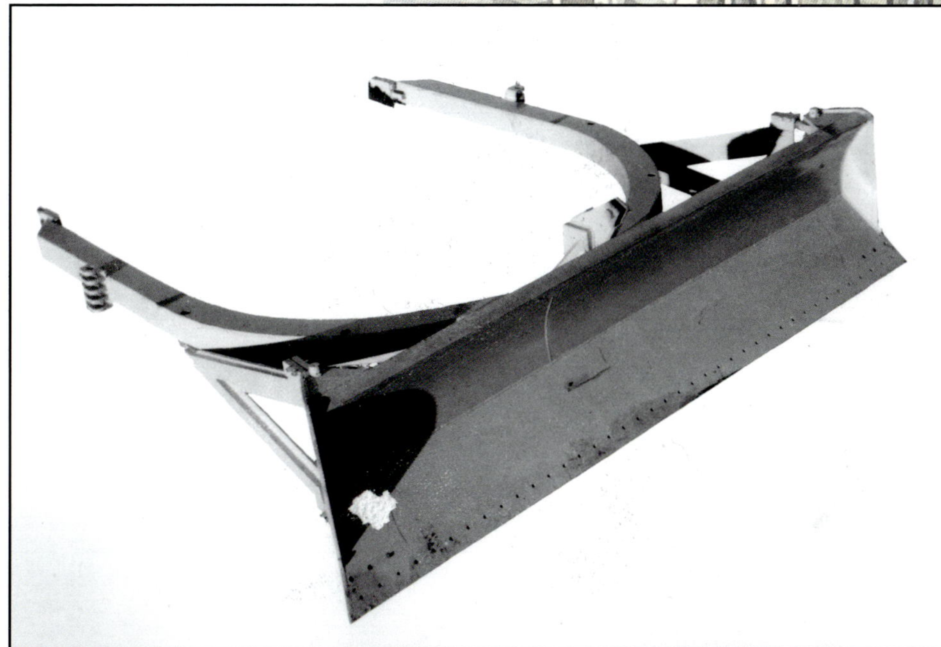

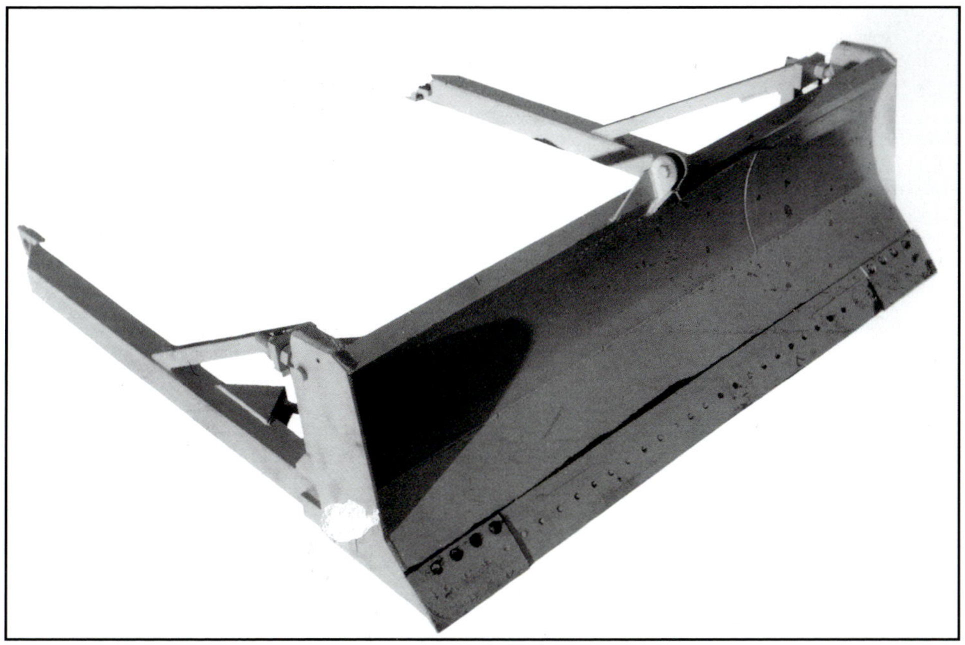

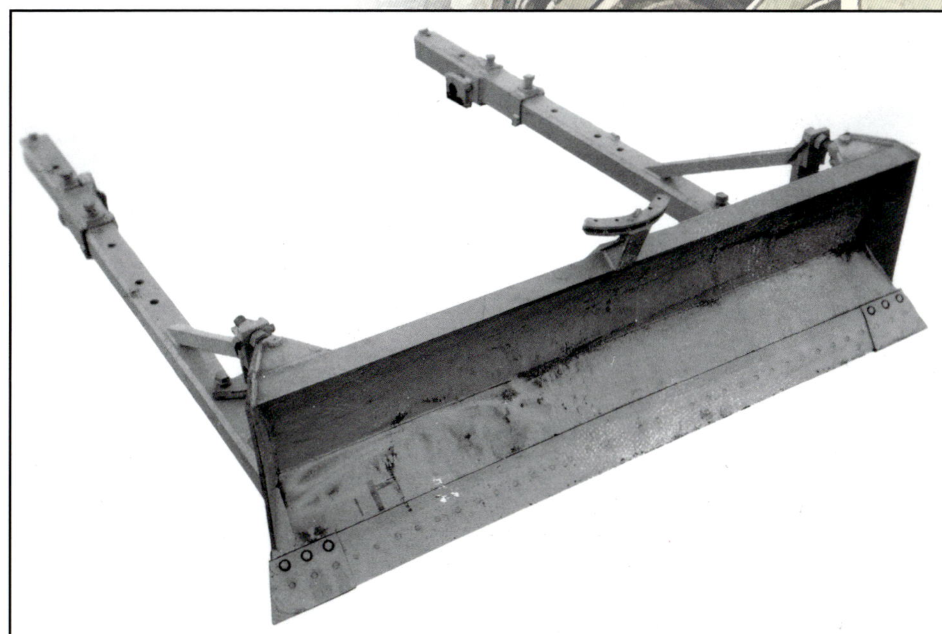

Several types of LeTourneau bulldozer kits were tested and used with Caterpillar tractors; four of them are illustrated here. At the upper left is the K-series bulldozer, introduced in 1938 for use on the Caterpillar D8 for the severest rock and quarry work. At the upper right is a CK-series angledozer for general-purpose work; a version, the WCK-7, was designed specifically for the D7 tractor. To the lower left is an XD-series bulldozer, designed for rapid knockdown and transport in fast-moving military operations. At the lower right is an AK-series angledozer, which, like the XD series, could be disassembled in three basic groups for transport. (R.G. LeTourneau)

U.S. Army students learn the fine points of lubricating a Caterpillar D7 while their instructor observes to the right. The student adjacent to the instructor is operating a volume pump, a manually operated canister for dispensing lubricant. A volume pump was a typical piece of equipment for a D7, being stowed in a holder on the right fender. (USAESHO) **Inset:** A front A-frame assembly for a bulldozer kit is viewed from the front, supported by a hoist chain. Braces are welded to the upper parts of the legs of the frame; at the bottoms of the braces are welded nonadjustable mounting brackets. At the center of the crossbar at the top of the A-frame is a housing for two sheaves, with braces. (R.G. LeTourneau)

This heavily weathered and hard-ridden D7 is equipped with a front-mounted single drum cable control for the LeTourneau dozer blade and a rear-mounted double drum Power Control Unit (PCU) for scrapers or other towed tools. The steel guard over the radiator grille is perforated with small, oblong openings. There was variation in these guards, with some having round perforations. (NARA)

A LeTourneau technician in white overalls instructs trainees on the use of the LeTourneau Model WCK-7 Tiltdozer. This dozer was characterized by a C-shaped yoke instead of individual pushbeams, and V-shaped sidearms linking the pushbeam to the dozer bowl. By repositioning the rear of the sidearm on brackets on the track rails, it was possible to vary the tilt of the dozer with reference to the longitudinal centerline of the tractor. (R.G. LeTourneau)

Students in a U.S. Army construction-equipment course adjust the tension of the left track of a D7. The trainee to the left is using a long-handled wrench to adjust a nut on the track-tensioning mechanism. The coil spring to the left of the two men to the right allowed the idler wheel to flex when the front of the track struck an obstacle. (USAESHO)

The army's construction equipment course covered all facets of maintenance and repair of a variety of tractors, including the D7. Here the tech sergeant instructor uses his pointer to indicate an issue concerning the bar and rope being used to pull a flywheel clutch assembly from a D7. A close-up view is provided of the top of the double-deck sheave assembly on top of the power-control unit. (USAESHO)

Students are receiving instruction in road-building at the Engineer School at Fort Belvoir. Two operators are practicing a maneuver called "tandem dozing" to move large quantities of material with two tractors operating as a single blade. Marked on the front of the hoods is "FB ES," for Fort Belvoir, Engineer School. Illegible white markings are on the ridge beam of the closer vehicle. (USAESHO)

At the Engineer School at Fort Belvoir, Virginia, a D7 heavy tractor pulls a towed-type Adams 124-S grader to improve a service road. The grader filled in the low spots and shaved off the high areas. The photo gives an idea of the angle of the bulldozer and pushbeam in the travel position. (USAESHO)

U.S. Army mechanics at Fort Belvoir are repairing the left steering clutch of a Caterpillar D7 equipped with a LaPlant-Choate Model B71 hydraulic bulldozer. The operator's seat and fuel tank have been removed to allow access to the drive train. The tractor in the background is a Caterpillar D8 with the two engine cylinder heads removed. (USACOE)

While a man stands in the road holding a survey level rod so a surveyor can check the grade during road-construction training at Fort Belvoir, the crew of a D7 take a break to the side. Hitched to the tractor is a LeTourneau Carryall scraper. In wet material, mud builds up on the track frames, as seen here. It requires cleaning at the end of the work shift to lessen excessive wear and damage in freezing weather. (USAESHO)

In the Cape Sudest area of New Guinea on 24 October 1943, two Caterpillar D7s are operating LeTourneau Carryall scrapers in a borrow pit, removing material for use on an access road at the Hanau Ammunition Dump. The D7 to the right has the curved push frame for an angledozer, but the dozer bowl-and-blade assembly is removed. The other D7 has a bulldozer attachment. Blade-and-bowl assemblies were removed when operating Carryalls to reduce weight and improve performance. (USACOE)

A massive concrete block is suspended from a LeTourneau M20 crane hitched to a Caterpillar D7 at the Engineer School, Fort Belvoir. This type of crane could lift up to 20 tons with the boom positioned in this manner. The hoist was operated by a cable from the D7's power-control unit. (USAESHO)

Top: The following sequence of photos documents the recovery of a crashed C-47 transport plane at the airport at Toccoa, Georgia, in October 1942. Here, a D7 heavy tractor, left, is operating a LeTourneau M20 crane, which is hoisting a main landing gear assembly from the crash scene. **Above:** The M20 crane is lifting the fuselage of the crashed C-47. The center section of the wing, along with the engines and nacelles, have been removed from the fuselage. Toccoa Municipal Airport was a site for training the 506th Parachute Infantry Regiment in 1942. When this plane went off the runway upon landing, the pilot claimed the runway was too short for C-47 operations, so the army moved the 506th's training operations to Fort Benning, Georgia. (R.G. LeTourneau, both)

Top: The M20 crane has recovered the center section of the C-47 wing, and the D7 heavy tractor is pulling the equipment from the crash area. **Above:** The Caterpillar D7 is drawing the LeTourneau M20 crane with the center section of the wing behind it. The tractor and the crane were assigned to the U.S. Army Engineers. (R.G. LeTourneau, both)

Top: Recovery operations continue of the crashed C-47 fuselage at Toccoa, Georgia. Despite the narrow tread between the crane's wheels and the seeming unlikelihood it could manage such a long, bulky load, the crane was up to the task. **Above:** The D7 and the hitched M20 crane have brought the C-47 fuselage from the crash site up to the level of the runway at Toccoa, where it will be hauled off for repair or salvage. The LeTourneau M20 crane was capable of lifting objects up to 40,000 pounds. (R.G. LeTourneau, both)

A LeTourneau M20 crane hitched to a D7 heavy tractor is positioned next to a beached Landing Craft, Mechanized (LCM) with markings of the 2nd Engineer Special Brigade, presumably in preparation for performing a repair or maintenance task. The boom of the M20 crane could be collapsed for travel or storage. When extended for operation, it was 28' 8" long, 9'3" wide, and 22' 6" high. It mounted 20-ply tires, either size 18-24 or 21-24. The M20 weighed between 7,600 and 8,400 pounds, depending on the type of tires. (R.G. LeTourneau)

Two tractors undergoing major maintenance at a base in England in February 1943, with a shop truck to the far right. The tracks have been uncoupled and are undergoing servicing, and men are working on the engine of the D7 to the right. Next to the left sprocket of that tractor is a special hydraulic puller for removing the sprockets. Tracks, rollers, and sprockets of crawler tractors were high-maintenance items; frequently requiring servicing, repair, and replacement. The two cylinder heads from the first tractor are on the work bench of the shop truck.To the far left is another D7 with the radiator guard lying on the ground next to it. Both of its tracks have been uncoupled. (NARA)

In New Guinea in 1943, a U.S. D7 equipped with a LeTourneau WCK-7 angledozer clears a road during the advance against Japanese forces. The operator is in the process of lifting the bulldozer, as earth is spilling out of it. Tractors such as the D7 were an essential asset in the formidable task of clearing roads and building bases in the thick jungles of the South Pacific and Western Pacific Areas. (R.G. LeTourneau)

The operator of a D7 fitted with a LeTourneau WCK-7 angledozer has his right hand on the right control lever of the power-control unit during a clearing operation on Bougainville on 30 November 1943. He was clearing an area along a continuation of the main supply line to the fighting front. The number 3 in a square is painted on the side of the radiator guard. (NARA)

At a tractor repair hut in an unidentified area in 1943, a soldier wearing the military lightweight tropical fiber sun helmet services the Dodge axle differential gears from the WC to the right. Behind him is a Caterpillar D7. Lying on the ground to the rear of the tractor is an International Harvester Model TD14 tractor engine. A power control unit is mounted on the rear of the tractor, and it is fitted for a LeTourneau bulldozer. (USAESHO)

At an airfield construction site in England on New Year's Day 1944, a Caterpillar D7 is towing a LeTourneau M20 crane with a concrete-finishing machine. The finishing machine had two wheels on each side that rolled on tracks as the machine moved along. It vibrated and helped smooth and finish fresh concrete. Here again, the M20 crane displayed its ability to carry wide loads in spite of its narrow wheel tread. (R.G. LeTourneau)

A Caterpillar D7 equipped with a LeTourneau WE-7 bulldozer performs maintenance work on Highway 6 in Italy, a key road from Rome to the fighting front at Monte Cassino. The photo was taken on 13 January 1944. The headlights on the front A-frame were adjustable, and the left one is turned downward. (NARA)

In a jungle in the South Pacific in January 1944, a D7 fitted with a LeTourneau CK-7 angledozer clears stumps and earth while widening and smoothing a newly constructed road. A partially visible name, possibly "Trail Blazer," and the code C-41 are painted on the top of the radiator guard. (R.G. LeTourneau)

A member of the U.S. Army Engineers repairs one of the V-shaped yokes of a WCK-7 angledozer for a D7 heavy tractor at a base at Milne Bay, New Guinea, on 2 February 1944. He was in the process of cutting through the rear end of the yoke. Welding on a reinforcing part would follow. (R.G. LeTourneau)

This Caterpillar D7 with an armored cab has markings that suggest it was assigned to Company B, 109th Engineer Battalion, with the Fifth Army in Italy. This vehicle was equipped with a LeTourneau R7 rear power-control unit and a WCK7 LeTourneau angledozer. This is one of the rare photos showing the blade tilted, left side down; the rear of the right sidearm is secured to the top setting of the stepped sidearm bracket to tilt the bulldozer blade. (NARA)

As the Allied armies began using crawler tractors more and more in combat situations, measures were taken to fit them with armored protection for the operators. This D7 with a hydraulically operated bulldozer assigned to the British 149th Assault Park Squadron is equipped with an armored enclosure fabricated in the U.K. The enclosure was of thin armor plates, with vision openings, ladder rungs on the sides, and no top. (IWM)

The same D7 shown in the preceding photo is viewed from the right rear. To the immediate front of the cab is the air precleaner on a pipe extension. At the rear is a Hyster D7N winch, above which is a toolbox. The Hyster D7N winch had a 13-inch-diameter drum, a line pull of 32,000 pounds on an empty drum and 22,000 pounds on a full drum, and weighed 2,450 pounds. (IWM)

A Caterpillar D7 fitted with armor for the operator and a LeTourneau WCK-7 angledozer pauses during obstacle-clearing operations near the Tanai River in northern Burma in late March 1944. The operator's armor appears to have been a field-fabricated affair of steel plates with no top. For self-defense two .30-caliber machine guns were provided on flexible mounts, and between them is the stock of a rifle. (R.G. LeTourneau)

In March 1944 a U.S. Army Caterpillar D7 bulldozer is being used to help English firefighters put out a farmyard fire by pushing straw around so the fire hoses can concentrate water on the smouldering materials. The operator of the D7, who has paused evidently to get further instructions from the fire marshal, is soaking wet from the spray. A faint recognition star is on the front of the radiator guard. (NARA)

In May 1944, U.S. Army Engineers employ a D7 with a LeTourneau angledozer in clearing ground for a new "super highway" supply route on Bougainville in the Solomon Islands. The D7 and angledozer could make short work of even the tangled trees and undergrowth of a jungle environment. "D-1" is painted on the side of the radiator guard. (R.G. LeTourneau)

In March 1944, Mt. Vesuvius in Italy began a series of major eruptions that caused much destruction, including to this U.S. Army Air Forces base. Here, a D7 heavy tractor with a LeTourneau WCK-7 angledozer cleans up cinders from the airfield in early June 1944. The still simmering volcano is visible in the distance. Over a foot of volcanic ash and cinders had covered the airfield. (R.G. LeTourneau)

Two men are at the controls of this Caterpillar D7 with a LeTourneau WCK-7 angledozer, engaged in cleanup work in a war-ravaged town in Italy around early June 1944. The bowl-and-blade assembly is in the fully raised position. The area probably was secure, as the operators aren't wearing helmets. (R.G. LeTourneau)

Privates William J. Shoemaker and Vinton W. Dove, combat engineers assigned to the 37th Engineer Combat Battalion, 5th Engineer Special Brigade, pose by a D7 with a WCK-7 angledozer after being awarded the Distinguished Service Cross for their heroic actions to clear obstructions under severe fire from the beachhead in Normandy on D-Day. Both men had formerly worked for construction companies. Stenciled on the side of the tool box is "DOVE—BILL." "HELLCAT" is stenciled on the ridge beam. (R.G. LeTourneau)

A Caterpillar D7 heavy tractor equipped with a LaPlant-Choate Model R76R angledozer is clearing debris in Isigny, France. This model of angledozer featured adjustable telescoping pushbeams for setting the angle of the dozer left or right, and matching jaws on the fronts of the pushbeams and on the rear of the dozer. Stenciled on the side of the fuel tank is the vehicle's weight, 33,238 pounds. (USAESHO)

Thirteen days after the D-Day landings in Normandy, a Caterpillar D7 with a LeTourneau WCK-7 angledozer clears a hedgerow in Normandy to prepare ground for the construction of an airfield. On the right fender is a row of U.S. five-gallon liquid containers and a single German Jerrycan in a storage rack, to the front of which is a toolbox mounted laterally on the fender. (R.G. LeTourneau)

The operator of a Caterpillar D7 heavy tractor has raised the bulldozer while clearing a hedgerow during construction of an airfield in Normandy on 24 June 1944. "Andre" is written in script on the brace of the front A-frame. A toolbox and what appears to be a fire extinguisher or a decontamination kit is on top of the hood. Four five-gallon liquid containers and a special high-volume grease gun are in a rack on the fender fabricated from metal strips. (NARA)

After shelling the French city of Valognes, on 22 June 1944 U.S. forces sent in engineers to clear rubble from the streets. Here, a Caterpillar D7 with a WCK-7 angledozer is pushing aside debris. Faintly visible in the dust to the left is another bulldozer. (NARA)

In another scene in Valognes during the Allied drive on Cherbourg, on 24 June 1944 a D7 bulldozer with "HARRIET" painted on the side of the hood stands by during cleanup operations. A rifle for the operator's self-defense is in a scabbard strapped to the left leg of the rear A-frame. To the side of the D7 is a crawler power shovel. (NARA)

G.I. labor is used to dig out a Caterpillar D7 with a R76R angledozer that is definitively mired in thick mud. The operator's station was surrounded on four sides by steel plate. In the background, a Diamond T 4-ton wrecker stands by to lend a hand in recovering the unfortunate tractor. (NARA)

This D7 is equipped for two missions. At the rear is a Hyster heavy-duty drag winch for heavy retrieval work. The bulldozer attachment at the front provides a universal tool for clearing, stripping and grading. A small single-cable drum Power Control Unit, the LeTourneau model FTD7 is mounted forward and is driven continuously from the engine crankshaft. The forward "live" PCU is used to raise and lower the dozer blade though the front A frame above the PCU. The operator control for the front PCU is the tall lever on the right side attached to a control shaft seen along the right side of the engine to the front mounted PCU. (USAESHO)

The same D7 displayed in the preceding photo is observed from the left side. The federal stock number, 2410-190-02327, is chalked on the push bar of the bulldozer, and the weight, 35,500 pounds, is stenciled on the track roller frame. A wooden packing crate for tractor parts is secured to the fender. Lifting eyes with the tops painted a light color are affixed to the front and the rear of the track roller frame. (USAESHO)

The same D7 is viewed from the rear, showing the rear of the Hyster winch. Below the winch is the drawbar assembly. Weight and dimensional data is stenciled on the rear of the fuel tank, and the federal stock number is chalked on it. Strapping to secure the parts crate to the fender extends around the fuel tank/seat and underneath the fender. (USAESHO)

A Tommy motions forward the operator of a British D7 clearing debris in a war-torn town. The tractor features an armored enclosure for the driver's station and engine compartment. The vehicle is equipped with a hydraulically powered bulldozer and a rear-mounted Hyster D7N winch. In addition to the toolbox mounted over the winch, several ammunition boxes are secured to the vehicle for stowing gear. (IWM)

A U.S. Army Caterpillar D7 is being used to push a British water ambulance into the English Channel on the French Coast on 10 July 1944. The D7 is fitted with a rear power-control unit and a LeTourneau WCK-7 angledozer. Water ambulances were small craft that evacuated wounded to England and returned to France with reinforcement troops. (R.G. LeTourneau)

A U.S. Army Caterpillar D7 with a LeTourneau angledozer operated by a rear-mounted power-control unit is in the act of recovering a Jeep from a bomb crater at St.-Colles, France, on 25 July 1944. "B24" is marked on the right rear of the fuel tank. The row of liquid containers on the rack on the fender are exclusively German Jerrycans: sturdy, serviceable containers that by this point in the war were being captured in large quantities and put to use in the Allied armies. There appears to be an overturned Sd.Kfz. 251 behind the D7. (NARA)

A U.S. Army D7 with a U.S. factory-made armored enclosure makes its way through rubble in Vire, France, in August 1944. The town had just been captured by Allied forces, and the push was on to reopen its streets so the advance could continue. The D7 had a hydraulic bulldozer and a rear-mounted LeTourneau power-control unit for operating towed machinery. The enclosure had a rear flap with a vision port and a side door. (NARA)

A group of G.I.s on the march take a break while a Caterpillar D7 with a LeTourneau WCK-7 angledozer fills in and smoothes a road in France in August 1944. The nickname "BOOGER KID" is stenciled on the "headache rack." Here again, Jerry cans have been commandeered for use on the vehicle. A toolbox is mounted on the hood. The number 45130 and "1-A" are stenciled on the top of the radiator guard. (NARA)

A Caterpillar D7 fills in a hole along the waterfront at Civitavechhia, Italy, in preparation for the construction of a new wharf in September 1944. The bulldozer is a LeTourneau WCK-7, with the telltale stepped bracket above and to the front of the trunnion for adjusting the bowl and blade's tilt. A small U.S. and Allied recognition star inside a circle, both of which were white, is stenciled on the toolbox on the left fender, and an even smaller star and circle are at the center of the ridge beam. (NARA)

The operator of a Caterpillar D7 with a LeTourneau WCK-7 angledozer has his hand on the right control of the power-control unit as he smoothes a road surface somewhere in Europe in World War II. A rearward-pointing service light is on the rear of each fender. On the rear plate of the LeTourneau power-control unit are markings for the 628th Engineer Company (LE), Third Army. (NARA)

Left: In Baume-les-Dames, France, in the 45th Infantry Division area of the Seventh Army, a Caterpillar D7 with a LeTourneau angledozer clears rubble from buildings destroyed by the retreating Germans in September 1944. On the rear plate of the LeTourneau power-control unit are markings for the 4th vehicle in the line of march of Company C, 120th Engineer Combat Regiment, 45th Infantry Division. A star and circle is stenciled on the side of the radiator guard. **Right:** In the Burma jungle on 20 October 1944, a Caterpillar D7 with a hydraulic bulldozer fells a tree during the construction of a logging trail. The D7 was assigned to the 797th Engineer Forestry Company, one of a number of such U.S. Army Engineer companies. These companies were tasked with producing timber and lumber in support of army operations. Each company consisted of a headquarters, a logging platoon, and a sawmilling platoon. (USAESHO, both)

A Caterpillar D7 with an angledozer, U.S.A. number 912257, assigned to the 834th Aviation Engineer Battalion, is clearing a field for the construction of an air base near St.-Trond, Belgium, 1 October 1944. On the side of the radiator guard is some risqué artwork. A late-type universal rifle bracket with a canvas cover is mounted on the right side of the air cleaner. (NARA)

During the October 1944 Battle of Aachen, Germany, U.S. Army troops along a railroad embankment await the order to advance while Ninth Air Force aircraft bomb nearby targets in the city. To the right, next to an M10 tank destroyer of the 634th Tank Destroyer Battalion, is a Caterpillar D7 tractor with an angledozer. (NARA)

Master Sergeant C. G. McGutcheon of the 1304th Engineer Construction Battalion pauses his D7 heavy tractor while a column of elephants carrying a survey party passes. McGutcheon was helping build the Ledo Road on 22 October 1944 as it neared its juncture with the Burma Road. His D7 is equipped with a hydraulic dozer; in the right background is another crawler tractor with a cable-operated bulldozer. (NARA)

A D7 bulldozer assigned to the Ninth Air Force's aviation engineers pushes a German dummy airplane out of the way at a base on the Cherbourg Peninsula of France around November 1944. This D7 was equipped with a cable-operated LeTourneau WCK-7 angle-dozer. Two wooden crates are strapped to the fender. (NARA)

At a site in the Pacific Theater, an engineer unit is mining coral to produce coral sand for use in the construction of an air base runway. To the right is a Caterpillar D7, scooping up coral with a cable-operated LeTourneau WCK-7 angledozer. (NARA)

At an asphalt plant operated by the 1889th Aviation Engineer Battalion on Guam on 21 October 1944, a D7 is pushing crushed rock toward a hopper. This plant was producing asphalt for the construction of Depot Field and Agana Air Base. The dump trucks in the background are delivering crushed rock to the plant. (NARA)

Near Loquignol, France, on 5 March 1945, a D7 assigned to the 1389th Engineer Forestry Company skids logs to a loading site while men in the background hoist logs on to a trailer for transport to a sawmill that was producing up to 3,000 board feet of lumber an hour for the construction of bridges across the Rhine. The D7 is fitted with a rear-mounted Hyster winch. Copious amounts of mud and dirt are packed between the sprocket, the top of the track roller frame, the upper run of track, and on top of the fender. (USAESHO)

A D7 with a LaPlant-Choate hydraulic angledozer skids a log to a loading area a few miles from Bitulok, Nueva Ecija, Philippine Islands, on 3 April 1945. A Hyster winch is mounted on the rear of the tractor, which has a hydraulic bulldozer. The Hyster winches offered far greater line pull than the LeTourneau power-control units. A heavy-duty winch was a must in logging operations of this type. (USAESHO)

A D7 assists in clearing a railroad track in a bombed-out town. The tractor has a LaPlant-Choate R-71 hydraulic angledozer. Raised headlights sport unique hoods with steel mesh guards to protect them from the frequent battering the front end of the tractor endured. (NARA)

One of the many crawler tractors the Allied armies put to work clearing and reconstructing devastated towns and cities, this Caterpillar D7 features a LaPlant-Choate R-71 hydraulic angledozer and a rear-mounted power-control unit. The object on the fender is the hydraulic reservoir for the angledozer's hydraulic controls. (NARA)

A Caterpillar D7 heavy tractor with a LeTourneau WCK-7 angledozer operated by a cable from a rear-mounted LeTourneau R-7 power-control unit clears a path through a destroyed neighborhood. To angle the dozer to discharge spoil to the left, the rear of the V-shaped left sidearm would be unpinned from its location shown here and pinned to the stepped sidearm bracket to the rear of the yoke. Either the lower left or right corner of the dozer could be tilted by pinning the sidearm to the higher steps. (NARA)

The operator of a D7 is demonstrating the operation of a LeTourneau M20 crane hitched to the drawbar of the tractor (the tractor engine is not running—a tin can rain cover is installed over the exhaust stack). The LeTourneau R-7 power-control unit on the rear of the D7 operated a cable that was routed through various sheaves to the hoist hook of the crane. Both of the cable drums of the power-control unit were employed in this operation: one to operate the hoist and one to work the boom. To the far left is the crane's tongue-sheave housing. (NARA)

A Caterpillar D7 prepares an earthen approach ramp to a landing site along the Rhine River on 6 April 1945. The tractor is equipped with an unusual, narrow bulldozer bowl and blade with push beams that are inboard of the tractor's track-roller frames. A Landing Craft, Mechanized (LCM) awaits loading with its ramp down in the background. (USAESHO)

During the Battle of the Ruhr Pocket in April 1945, a Caterpillar D7 heavy tractor is working on the construction of an approach for a pontoon bridge on the Dortmund-Ems Canal between Essen and Bottrap, Germany. Trucks hauled in fill for this operation, and it took three bulldozers an hour to construct the approach. A heavy smokescreen had been laid to mask this operation from the Germans. (R.G. LeTourneau)

Top: A pair of crawler tractors, including a D7 in the foreground, are towing LeTourneau Carryalls for use in construction work. Marked on the side of the radiator guard of the D7 in the foreground is "250-E," probably a reference to an engineer regiment or battalion. (USACOE) **Above left:** A D7 with a LaPlant-Choate angledozer has become mired in a muddy slough along a road. The men in front of the dozer are shoving planks in front of the tracks in an attempt to give them some purchase. In the end, it may have been necessary to call in a wrecker. (Patton Museum) **Above right:** An armored Caterpillar D7 clears debris in a European town. "Ol' Mud An' Guts," a play on Gen. George S. Patton's nickname, "Old Blood and Guts," is painted on the cab. Two angle-iron splashguards are on the angled front of the cab below the vision port. (Patton Museum)

In the final days of the war in Europe, a D7 with a LeTourneau WCK-7 angledozer pushes aside log barricades the Germans left in the town of Heppendorf, Germany, to slow down the American advance. Remnants of the barricade are to the left. (NARA)

In what appears to have been a panzer graveyard, a Caterpillar D7 with a LeTourneau WCK-7 angledozer and R-7 power-control unit is preparing to drag away the upside-down upper hull and turret of a destroyed Pz.Kpfw IV tank. To the right is the idler and track of an upended hull, and in the center background is another Pz.Kpfw. IV awaiting disposal. Worthy of notice is an empty universal rifle rack on the ridge beam and the fact that the headlights on the forward A-frame have been turned to the rear. (NARA)

In this photo of a D7 working on a road in hilly country, a universal rifle rack installation on the ridge beam is shown with an M1 Garand rifle and sling secured in it. The dozer unit is a LeTourneau WCK-7 pushdozer powered by a cable. In the background is another crawler tractor with a WCK-7 and a Hyster heavy-duty winch mounted on the rear. (USACOE)

Two soldiers hold communications lines out of the way while a D7 with an angledozer pushes an abandoned German Panther tank to the side of the road. Although not visible from this angle, an FTD-7 power-control unit to the front of the radiator guard operated the angledozer, and because the unit was front-mounted, a rear A-frame was not necessary. A Hyster D7N winch is mounted on the rear of the tractor. (NARA)

Members of an engineer unit who apparently were expecting to be operating soon under combat conditions don flak vests next to and on top of a Caterpillar D7 with a LeTourneau WCK-7 angledozer. A clear view is available of the eyebolts where the sidearm joins the dozer bowl. By turning nuts, these eyebolts were adjustable for setting the vertical tilt of the bowl and blade to increase or decrease digging action. (USACOE)

Along Highway 5 on Okinawa in May 1945, a Caterpillar D7 constructs a bypass around a bridge blown by the Japanese to hinder the American advance. The dozer was a LeTourneau WCK-7, as indicated by the stepped sidearm bracket and the V-shaped sidearm partially visible. A five-gallon liquid container and a grease pump are stored on the fender. As is often seen, the operator has a spare coil of wire rope or cable on the rear A-frame. Control cable was a high wear item. Breaking a dozer lift cable meant that the operator could only move in reverse dragging the un-liftable dozer bowl to a safe location to install a new cable. (NARA)

At Bacolod Airstrip, Negros, Philippine Islands, on 18 May 1945, a Caterpillar D7 is recovering a C-47 cargo plane that went into a drainage ditch. The tractor was assigned to the engineers of the 40th Division. With this type of configuration of a rear, twin-drum power-control unit but no dozer or A-frames, the tractor was set up for operating a LeTourneau Carryall scraper. (NARA)

Top left: Mechanics service a D7 with an armored cab and a LeTourneau XD7 bulldozer somewhere in the Pacific. The inner surface of the side door is visible. It appears to be painted white or a light color and has a locking handle. Aft of the door is a vision slit. **Top right:** The same D7 is viewed from a different angle. The front of the armored cab has two vision ports with hinged covers. A vehicular toolbox is on the roof of the cab. Above the man to the left is an exhaust stack extension apparently made from soldered tin cans. **Above left:** On the same D7, the rear window of the armored cab has a flap and a wire-mesh grille. The sign on the post to the left of the tractor reads "HEAVY MAINT[ENANCE]. SECTION." On the rear of the tractor is a LeTourneau double-deck power-control unit. **Above right:** In mountainous terrain in the Pacific Theater, a D7 equipped with an A-frame assembly but no dozer is towing a road grader. As was the case almost anywhere a D7 operated in soft soil, a large heap of dirt has accumulated on top of the track roller frame. (USACOE, all)

Top: Engineers take a pause during a construction project in the Pacific Theater. In the foreground is a Caterpillar D7 with an A-frame assembly and rear-mounted power-control unit. The box-shaped object jutting from the front of the chassis housed the LeTourneau Model AD7 single-drum power-control unit for a bulldozer or angledozer attachment. Note the cable running up to the sheave assembly on the front A-frame. A trunnion for an angledozer or bulldozer is faintly visible on the side of the track roller frame. Hitched behind the D7 and powered by cables from its power-control unit is a LeTourneau Carryall scraper. Visible in the center background is another D7. **Above:** At a road-construction or repair site in the Pacific Theater, a Caterpillar D7 with a LeTourneau XD7 bulldozer is pulling a grader. Normally, graders were used in the finishing part of a road job, but here it is operating on a surface that is still quite rough. The operator of the D7 has hung his goggles and several small stowage bags from the ridge beam. (USACOE, both)

A Caterpillar D7 fitted with a LeTourneau angledozer is slot dozing for a bomb shelter and medical-aid station on the beachhead at Lingayen, Philippine Islands, in May 1945. The dozer has cut a shelter trench as deep as the height of the fenders. Track prints are visible in the sand to the rear of the vehicle. (NARA)

A D7 is towing a U.S. Navy Construction Battalion "Sea Bees" GMC CCKW shop truck to shore from a landing ship, tank (LST) on Blue Beach, Vieques Island, off the Puerto Rican coast during Operation Portrex in February or March 1950. The bulldozer is the LeTourneau XD7, and the tractor has a rear-mounted Hyster winch. At the rear of the shaft running along the fender and the engine is a lever for controlling the front power-control unit. Combat packs are stowed on the A-frame. (USAESHO)

A Caterpillar D7 serving with an aviation engineer unit—possibly the 818th Aviation Engineer Battalion, judging from the markings—is parked at an airfield with a Republic P-47 in the background. The tractor has a single-drum front cable control for a bulldozer. The dozer was removed when the D7 was operating a Carryall scraper, part of which can be detected to the far right. The operator is in the normal position for operating a scraper, perched on the left armrest to allow visibility both fore and aft. (USACOE)

Toward the beginning of the Korean War, a Caterpillar D7 is backing onto a Landing Craft, Tank (LST) at the Po-Hang Basin near Yongdok, Republic of Korea, on 28 July 1950. On the landing craft behind the D7 is a LeTourneau Carryall scraper, a piece of heavy-construction equipment that was cable-powered by the towing tractor's power-control unit. (NARA)

Several crawler tractors are undergoing work in the tractor shop 54th Engineers Field Maintenance Company, Eighth Army, in the Republic of Korea on 3 September 1953. The nearest one, Caterpillar D8 U.S.A. number 9202764, has a Hyster D8L winch, which was specific to the D8. The next tractor is a postwar Caterpillar D7 fitted with a Caterpillar Number 25 rear cable control unit. The Cat Cable Control replaced the LeTourneau-supplied rear PCU. (USAESHO)

A D7 bulldozer is shoving rocks down a wooden chute to a rock crusher operated by Company A, 4th Engineer Combat Battalion, First U.S. Corps, in the Republic of Korea on 24 November 1953. There is a pause in the operation, and several engineers are standing on the D7's track. (USAESHO)

Left: Two 6-ton Tractor Cranes M4 working side by side are hoisting a large, wooden crate containing an M3A1 half-track from or to a railroad flatcar in England in 1943. The 6-ton Tractor Crane M4 comprised a Cardwell Model C crane of 12,000 pounds capacity mounted on a Caterpillar D7 tractor. The crane was cable operated by a winch housed inside a metal hood at the rear. (MHI) **Right:** The same pair of 6-ton Tractor Cranes M4 shown in the preceding photograph continues to lift a wooden crate holding an M3A1 half-track. For this crawler crane, the tractor had modified suspension with an extended track roller frames to hold seven rollers instead of the usual five on each side. The box-shaped object to the rear of the boom was a counterweight. (MHI)

Top left: Photographed in the 1950s, this D7 has a Caterpillar Number 25 rear-mounted power-control unit, a Caterpillar No. 24 front power-control unit, and a Caterpillar C-frame for an Caterpillar angledozer. **Top right:** The same D7 shown in the preceding photo is viewed from the right. The front power-control unit is barely visible to the front of the grille. A parts crate is on the fender. **Above left:** In a rear view of the same D7, U.S.A. number 9202931, are the rear Caterpillar power-control unit along with a shipping placard, destination Fort Richardson, Alaska. **Above right:** This D7 was nearly new when photographed in the 1950s. It has a Straight S bulldozer and front power-control unit, connected by the long shaft to the control lever to the right of the operator's seat. (USAESHO, all)

By the late 1940s, some U.S. Army D7 heavy tractors were equipped with accessories for performing pipeline-construction work. This example, U.S.A. number 9211924, is equipped with the fixtures for mounting a cable-operated Trackson Model MD7 Pipelayer, the boom of which is not installed. Because of the presence of the Pipelayer's crossbeam at the front of the operator's station, the oil-bath air cleaner has been moved to the side of the engine. (USAESHO)

Top left: The boom of the MD7 was attached to the brackets on the left track roller frame. Between them is the vertical frame for sheaves for cable operation of the boom. **Top right:** On the right side of the MD7 Pipelayer at the center of the D7 is a counterweight assembly, to balance the vehicle when the boom was lifting heavy pipes. **Above left:** The MD7 viewed from the front. A heavy-duty frame was required to support the weight of the counterweight assembly, the operating machinery and the pipe-laying boom. **Above right:** The Trackson MD7 Pipelayer installed on a Caterpillar D7 is seen from the rear. The transmission provided power for it. A parts crate was stowed on the operator's seat. (USAESHO, all)

A D7 beach tractor has weights attached to the hoist cable at the rear of the crane, for performing lifting tests. The beach tractor was developed for the Engineer Amphibian Command by Donald Roebling, who also developed the Roebling Alligator, prototype of the LVT amtracs, and who was the grandson of Washington Roebling, builder of the Brooklyn Bridge. (USACOE)

This Caterpillar D7, U.S.A. number 969473, probably a 3T-series machine, was modified as a beach tractor. It had an American-type armored cab with a side door and featured a tubular crane mounted on the rear, for unloading landing craft; a rear-mounted Hyster D7N heavy-duty drag winch, for pulling vehicles from the beach; and a unique heavy-duty, narrow bulldozer, for constructing exit lanes from the beach. (USAESHO)

Left: D7 beach tractor U.S.A. number 969473 was equipped with a bulldozer blade and bowl narrow enough to allow the tractor to board small landing craft. Apparently these were about the same width as, or possibly slightly narrower than, the outside measurement of the track width. The push beams were positioned inboard of the tracks and the track roller frames: here again, to keep the overall width of the vehicle as narrow as possible. At the time, virtually all bulldozer push beams were mounted on the outboard sides of track roller frames.

Right: The D7 beach tractor is observed from the rear. Dominating the view is the Hyster Model D7N heavy-duty drag winch. The beach tractor had a fairlead system under the chassis for directing the winch cable to the front of the tractor. Above the winch is the rear of the armored enclosure for the operator and the mounting for the crane, which had a short, "wishbone" type boom. The crane was deployed through a rack-and-pinion gear at the rear. (USAESHO, both)

The D7 beach tractor's bulldozer was cable operated by the standard front power-control unit; thus, only a front A-frame was necessary. At the top of the D7 beach tractor's crane boom, seen in the travel position, was an unusual dual sheave assembly, with multiple large and small sheaves in evidence. On the armored enclosure, an open vision port is at the front, and the right side door is in view. To the rear is the right rack-and-pinion gear and handwheel for the crane; a similar apparatus is on the left side of the tractor. (USAESHO)

In a head-on view of a Caterpillar D7, the tractor is equipped with an angledozer with its left side tilted down, engaged in performing a side-hill cut. Worthy of notice are the tall, straight exhaust stack and the absence of a muffler. This was apparently a test vehicle, marked on the top of the radiator guard "ASF-E BOARD 252." (USACOE)

This restored, fully operational Caterpillar D7 heavy tractor was in the collection of the late Kevin Kronlund. This tractor was produced in 1943 and is a 7M-Series machine: D7s are referred to by series, and series are designated by serial-number prefix. The first-series D7, produced from about 1940 to 1944, was assigned serial numbers 7M0002 to 7M9999. Here, Jeff Rowsam is at the driver's controls. The tractor is equipped with a LeTourneau XD7 bulldozer operated by a cable from a rear-mounted LeTourneau R7 power-control unit. (DD)

Top left: In a view of the front end of the D7 heavy tractor from the right, details of the track are in the foreground. Rising above the hood is the Diesel exhaust stack. To the right is the A-frame of the LeTourneau XD7 bulldozer, by which the operating cable was routed to the dozer. (DD) **Top right:** The engine is viewed from the right. To the front of the engine is the fan. Grouped along the side, front to rear, are the governor housing, fuel transfer pump, fuel-injection pumps, and the two oil filters. Above and to the rear of the engine is the air cleaner. (DD) **Above left:** The governor housing, fuel transfer pump, and fuel-injection pumps are viewed close-up. At the lower center, a fuel line is connected to the fuel-transfer pump. Access panels with hex screws are provided to adjust the fuel injection pump timing. Four fuel injection lines lead to each of the injection nozzles in the two cylinder heads at the top left. (PS) **Above right:** The two vertical canisters at the center are the oil filters. Inside each oil filter housing were two filter elements: a washable outer element and a replaceable inner element. Filters were to be cleaned and sludge drained from them at each oil change. (PS)

Top left: At the center of the photo, molded into the front of the water-return line toward the upper right front of the engine block, is the housing for the temperature regulator of the cooling system. The part number, 6B7276, is cast in raised figures on the side of the housing. (DD) **Top right:** In a view of the hood from the right, to the left is the air cleaner, topped with the precleaner with a glass jar attached to its underside. To the side of the air cleaner is a rifle in a universal bracket. Atop the hood is a vehicular first-aid kit, a common modification. (DD) **Above left:** In a view of the right side of the engine, to the side of the sheet metal bodywork to the right is a steel plate that acts as a mounting for the right side of the A-frame of the bulldozer assembly. The bottom of the plate is bolted to the chassis frame. (DD) **Above right:** The driver's station of the Kevin Kronlund D7 is viewed from the front right, showing the spacious, cushioned seat with arm rests and the control levers. To the right are two five-gallon liquid containers, to the rear of which is the hand-operated volume pump. (PS)

Left: Mounted in a holder on the right fender and secured in place with a webbing strap is a special portable hand-operated large-volume grease compressor pump, for use in daily the lubrication of the track rollers and idlers of the vehicle tractor. This volume-compressor pump was manufactured by Alemite, as part number 6521, and was issued with the tractor from Caterpillar. It features a canister, a pump handle with a locking lever, and a flexible hose that is fitted on the end with a special adapter to connect to the large button-type grease fittings on each track roller. **Top right:** As viewed from the right side of the seat, an instrument cluster is to the right of the air filter. Two steering levers, one for left and one for right were used to disengage a multiple disk steering clutch on each side to achieve directional control. A separate brake pedal to each steering clutch could be used with the steering clutch for sharper turns. The circular fixture to the front of the seat is the fuel-pump and governor control. (DD) **Above right:** The fuel-pump and governor control with its operating lever and knob is seen from a different angle. The right brake pedal is seen to the right rear of the air cleaner. The driver's seat was positioned in front of and on the fuel tank, an L-shaped structure as viewed from the side. (DD)

Top left: Fitted to this D7 is a LeTourneau XD7 bulldozer, as observed from the right side. The bottom rear of each side plate of the bulldozer was secured with a pin to a pushbeam. The top of both side plates were fastened with an eyebolt to the top of a V-shaped brace. (PS) **Top right:** The LeTourneau XD7 Bulldozer was fitted with a large, threaded stud with dual hex nuts that were turned forward or to the rear to adjust the tilt of the bulldozer bowl and blade, resulting in different digging properties. This eyebolt is on the right side of the dozer. (DD) **Above left:** Facing the rear of the right side of the dozer, the brace for the right eyebolt is to the right of the photo. At the lower center is a brace assembly. There are two, and this is the right one. It is fastened with a pin to a bracket welded to the bottom of the bowl of the dozer. (DD) **Above right:** Pinned to a bracket welded to the top center of the dozer bowl is the hoist sheave assembly. It contains dual, side-by-side sheaves. The sheave assembly acted as a force multiplier for the operating cable, routed from the power-control unit at the rear. (DD)

Top left: The front A-frame of the LeTourneau XD7 bulldozer is seen from the right. On each side of the hoist sheave is a headlight. Attached to the crossmember of the A-frame is the ridge beam. **Top right:** At the front end of the D7's suspension is the right idler wheel. It is mounted on adjustable bearings to permit proper tensioning of the track. The horizontal beam along each side of the tractor that supports the suspension and the vehicle is the track roller frame. **Above left:** Two track-return rollers on each side of the D7 support the upper span of the track. Technically, these were referred to as track-carrier rollers. The roller is mounted on a shaft attached to the upper end of a horizontal arm, visible from this angle. **Above right:** The front right track-return roller assembly is viewed from the rear, showing the design of the arm supporting the roller shaft. Each roller was a single part, fastened to the shaft with a cap and a washer. To the far right are the right idler wheel and its yoke. The special button-type grease fitting is seen on the roller end cover. (DD, all)

Top left: The tracks are composed of overlapping links, held in a chain with pins and bushings. Each track assembly had one master pin, which was split on each end and held in place by a tapered plug. To remove the track, it was necessary to locate and remove the master pin. (PS) **Top right:** A view through the space between the right track roller frame and the side of the body. To the right is a track-return roller arm, and faintly visible at the center, under the body, is the auxiliary spring, a leaf spring mounted underneath the larger equalizer spring. (DD) **Above left:** The right side of the equalizer spring is in the left foreground. The equalizer spring transfers the weight of the front of the body to the track roller frame. Its auxiliary spring holds the equalizer spring in place in its saddle and allows it to oscillate with track movement. (DD) **Above right:** To the rear of the track roller frame is the sprocket, a single wheel with teeth that engage the bushings at the center of the track. The sprocket's outboard bearing is mounted on the track roller frame. To the right is the trunnion for the push beam of the dozer. (DD)

In a view of the Kronlund D7 heavy tractor, a good look is offered of the large-volume compressor for dispensing grease stowed on the right fender, with a toolbox and a five-gallon liquid container to the front of it. To the rear of the seat back is the upper part of the fuel tank, with the filler cap located at the center of the top of the tank. Hovering over the top of the tank and seat is the right control lever for the power-control unit. (Jeff Rowsam)

The tall pipe on the hood is the engine exhaust. It had a hinged cap that blew open when the engine was running. Sometimes these got broken or detached, so a tin can often was placed over the top when the engine wasn't running. The smaller pipe shaped like an inverted J on the hood is the starting engine exhaust. To the rear of the engine exhaust is a vehicular first-aid kit, to the rear of which is the air cleaner, with the centrifugal precleaner on top, with a clear glass jar for collecting debris. (Jeff Rowsam)

The Caterpillar D7 Heavy Tractor is viewed from the right rear, showing the positions of the rear A-frame and the LeTourneau R7 power-control unit. Raising and lowering of the dozer was accomplished through the power-control unit, a dual-drum winch from which a wire-rope cable passed up, through the ridge beam, and down to the hoist-sheave assembly on the top of the dozer bowl. The power-control unit also was used for powering specific accessories, such as the LeTourneau M20 crane, but was not used for general winching operations. (DD)

Left: The arrangement of the rear A-frame, ridge beam, and power-control unit are viewed from the rear. The rear A-frame was asymmetrical, and skewed toward the right, in order to position the rear of the ridge beam over the power-control unit's right drum, which fed cable to a sheave housed inside the rear of the ridge beam. Below the power-control unit is the drawbar with a tow pintle at the center. **Top right:** The space between the rear of the fuel tank and the power-control unit is seen from the right. At the bottom center is the attachment point of the bottom of the right leg of the rear A-frame to the rear of the chassis frame. A grab handle is on the side of the fuel tank. **Above right:** The rear of the fuel tank is marked with a star. To the sides are the legs of the rear A-frame. To the right of center is the operating lever for the right cable drum of the power-control unit. Below the fuel tank are access covers for the steering clutch assemblies. (DD, all)

Top left: The power-control unit received power through a spline shaft mated to the upper transmission shaft. Not present is a double-deck sheave housing that mounted atop the unit and permitted the cables to be directed in various ways for specific tasks. **Top right:** The cable used in the LeTourneau R-series power-control units was the Tournarope wire rope or a comparable product. It was 1/2 inch in diameter, 6 x 19 (6 strands, 19 wires per strand) Lang lay (wound in the same direction), made of improved plow steel. **Above left:** The area at the rear of the D7 below the power-control unit is viewed. At the bottom is the drawbar assembly. The bracket supporting the right side of the drawbar assembly is to the right. At the upper right is the lower part of the power-control unit. **Above right:** The left drawbar bracket is viewed close-up. At the bottom is the drawbar. To the far left is the left final-drive assembly. At the center is a cap with four hex screws and a grease fitting for retaining the track roller frame; the corresponding right cap is to the far right. (DD, all)

Left: The LeTourneau R-series power-control unit and the double-deck sheave are displayed, along with the rear A-frame and the drawbar assembly. To the front of the power-control unit are a pioneer tool rack and the operating levers for the unit. Normally, the levers were in the neutral position. To spool cable onto the drums, the levers were pushed to outboard; to release cable from the spools, the levers were pushed inboard.

Right: The double-deck sheave assembly atop a LeTourneau R7 power-control unit is observed from the left side. The sheaves pivoted, so it was necessary to lubricate the pivot points, as well as the sheave bearings, after every eight hours of use using general-purpose grease. Care was to be taken not to over lubricate the sheave assembly, to prevent dripping grease from affecting the clutch and brake facings. (PS, both)

Left: The space between the fuel tank/operator's seat and the power-control unit is portrayed. A C-shaped grab handle is welded to the side of the fuel tank. The attachment point of the left leg of the rear A-frame is to the left of the coil spring on the power-control unit. The control levers of the power-control unit are shaped like inverted Ls, with the grab handles to the upper front. (PS) **Top right:** On each side of the rear of the tapered rear plate of the power-control unit is a small clamp plate that retains the left and the right cable-drum shafts. These clamps are secured to the rear plate with hex screws. Also in view at the bottom is the drawbar assembly. (PS) **Above right:** At the rear of the drawbar is a yoke and pin for hitching a towed vehicle; the front of the drawbar pivots in a bracket. Supporting the drawbar is a curved rail with holes bored through it so the drawbar can be positioned at different angles and secured with a pin. (DD)

Top left: The left cable drum of the power-control unit is viewed from the side. Surrounding much of the drum is a cable guard. Toward the left, the square nut is part of the plug for the gearbox oil filler; below it is a plug for checking the level of the oil in the gearbox. (DD) **Top right:** An overall view of the rear of the Kronlund Caterpillar D7 includes details of the tracks, the rear of the fuel tank, the rear of the power-control unit, the left and right final drives, and the drawbar. The caps for the roller track rails are visible above the drawbar. (Jeff Rowsam) **Above left:** In a view from the left fender of the D7 heavy tractor, in the foreground is an M1 Garand rifle secured in a universal rifle bracket with a cover over the muzzle. To the far right is the oil-bath air cleaner, with the glass jar for collecting debris below the precleaner. (Jeff Rowsam) **Above right:** A lamp is mounted on the rear of the left fender, facing to the rear. The sheet-metal fenders have curved outer corners, and their edges are turned down to add strength to the assemblies. Above the lamp on the side of the fuel tank is a C-shaped grab handle. (DD)

A Caterpillar D7 heavy tractor equipped with a LeTourneau power-control unit with double-deck sheave assembly and LeTourneau bulldozer kit is viewed from the left rear. The dip of the upper run of the track between the tops of the idler, the return rollers, and the sprocket is apparent. Atop the fuel tank to the left of the left leg of the rear A-frame is the fuel filler cap. (PS)

Operator Jeff Rowsam puts Kevin Kronlund's Caterpillar D7 through its paces. Lashed to the ridge beam with webbing straps is a post-hole shovel for cleaning mud from the track frames. An M1 Garand rifle is secured to a universal rifle bracket to the rear of the engine. When the bulldozer is raised in the travel position, there are no travel locks to support its weight; the hoist cable does the job. (DD)

Top left: At the operator's station, seen from the left, the nearest lever is the main flywheel-clutch lever. In front of the center of the seat, on each side of the fuel-pump and governor control (or throttle) are the steering clutch levers, to the front of which is the air cleaner. To the far left is a universal rifle rack, because a D7 operator never knew when he might have to fight for his life. (DD) **Above left:** Above the center of the seat back is the control for the right drum of the power-control unit. On the floor are the left and right brake for each steering clutch. (PS) **Right:** The operator's control pedals and levers are seen close-up from the left. The nearest pedal was for the steering clutch. The fuel-pump and governor control, also referred to as the throttle, is the round object with a lever on top of the tubular pedestal. The two levers on the far right side are the transmission gear selector and the forward/reverse shifter. In the background are a five-gallon liquid container and the large-volume grease compressor. (PS)

Top left: At the left side of the D7 Diesel engine is the 2-cylinder gasoline starting engine. Four hand controls levers are used for starting the Diesel engine. They are the compression release, the starting engine transmission hi-lo shifter, the starting engine pinion lever and the clutch lever. The vertical shaft is for cranking the starting engine through the hood opening with a manual hand crank. (PS) **Top right:** This D7 Diesel engine was the military version of the Caterpillar D8800 industrial engine. It was a four-cylinder, naturally aspirated, pre-combustion chambered, low-speed design rated at 900 or 1,000 rpm and 80 horsepower. It had an 8-inch stroke. (DD) **Above left:** In a view of the left side of the D7 engine from the front, to the rear of the A-frame brace is the carburetor and air cleaner for the starting engine. Behind the carburetor is the intake manifold. Situated to the right of the right point of the star is the water pump. (PS) **Above right:** The design of the front A-frame of the Kronlund D7 heavy tractor is viewed close-up. The legs are of welded box-girder construction, with beveled bottoms. "PRESTONE 42" on the radiator guard on a WW2 D7 meant the radiator received antifreeze in 1942. (Jeff Rowsam)

Top left: At the midpoint of each track roller frame is the pivot point for the push beams, referred to as the trunnion. Above the trunnion, on top of the track roller frame, is a guard for the idler spring. Also available, but not installed here, was a guard for the sprocket bearing. **Top right:** The left idler wheel and yoke are viewed close-up. Details of track construction also are shown. Each track shoe is secured with bolts and nuts to two track links; each of these assemblies is connected to the next one with a pin through a collar and a bushing.

Above left: The front A-frame of the LeTourneau bulldozer kit is viewed from the left rear. The sheave housing at the center top is braced with a diagonal tube on each side, the bottoms of which are welded to the inner sides of the A-frame. Headlights also are in view. **Above right:** The sheave housing, top of the front A-frame, and headlights are observed from the front. On the side of the sheave housing is a pocket and a wedge that secure the end of the bulldozer operating cable. On the front of each headlight lens is a steel-mesh guard. (PS, all)

Left: Affixed to the upper center of the bulldozer bowl is a dual-sheave assembly. The routing of the operating cable of the bulldozer through multiple sheaves had the effect of multiplying the force of the power-control unit, as opposed to the much greater force that would be required if the cable were routed directly from the power-control unit to the bulldozer bowl. The bottom of the dual sheave housing is secured with a steel pin to a bracket on the bowl. (PS) **Top right:** The threaded stud and nut for adjusting the tilt of the bulldozer bowl and blade are in view at the top of the left brace. A similar nut is out of view on the front side of the bracket. Both nuts were moved forward or back and then tightened to set the tilt. (DD) **Above right:** As seen from behind the right rear of the bulldozer bowl, this is the left of two braces running from the inner side of the push beam to the lower rear of the bulldozer bowl. A bracket welded to the rear of the bowl holds the front end of the brace. (PS)

The bulldozer assembly is raised during a demonstration of the D7's capabilities. The blade, or cutting edge, at the bottom front of the bulldozer bowl is a three-piece assembly, with two shorter, taller sections to the side, which usually wore out first, and a longer center section that was lower in height. All three sections were removable and replaceable. (DD)

Jeff Rowsam, an expert on Caterpillar D7s and the technical consultant for this book, demonstrates a bulldozing operation on Kevin Kronlund's D7. The operator of a fixed bulldozer, as opposed to an angledozer, required exceptional skill to ditch, backfill, or cast material in an effective, efficient manner. (DD)